Amazon Fire TV User Manual: Guide to Unleash your Streaming Media Device

By Shelby Johnson

www.techmediasource.com

Disclaimer:

This eBook is an unofficial guide for using the Amazon Fire TV product and is not meant to replace any official documentation that came with the device or accessories. The information in this guide is meant as recommendations and suggestions, but the author bears no responsibility for any issues arising from improper use of the Amazon Fire TV. The owner of the device is responsible for taking all necessary precautions and measures with their device.

Contents

Introduction

Amazon created the Fire TV device to change how we watch television. Over the last 10 years, there has been a revolution and evolution in the way that people consume media. With the development of streaming technology, smart phones, tablets and other mobile devices the way we watch media has changed. Streaming networks like Roku, Netflix along with DVRs and on demand cable have created an "always ready when you are" model of watching TV.

No longer do people ink in a set time to watch their favorite shows. Now they stream them when they are ready to watch them. In many cases people have chosen to completely cut traditional cable TV service altogether, and they are saving money while being happier at their TV choices to boot.

With this new device, Amazon has made inroads into the market because they have produced an exceptional streaming device that also has bonus real gaming potential. Amazon has always been creative with its applications and production of products, and Fire TV does not disappoint. There are a range of applications available for those users that want something effective and affordable. Amazon Fire TV is home to numerous high-quality applications that are constantly being updated to produce superior results.

I purchased my Fire TV the day it came out, and I put together this guide to help you get the most out of your device. Let's take an in depth look at the Amazon Fire TV straight out of the box.

Note: *Amazon Fire TV is so new that not everything is functional with the device at this time. Amazon Fire TV FreeTime will not be active until May 2014.*

Amazon Fire TV vs. Apple TV vs. Chromecast vs. Roku

Amazon Fire TV has three main competitors in the live streaming industry - the Apple TV, the Google Chromecast, and the Roku streaming media player. Although all four devices allow you to stream movies, television programming, and music, Amazon tops all three devices because it has things that the others don't: Superior technology and real gaming potential.

The Roku comes in a number of tiered options, allowing you to choose from a simple version with standard broadcasting capabilities, to one with HD formatting and a microSD card. Amazon Fire TV, Apple TV, and the high end Roku set top boxes all cost the same amount, and they all provide similar amenities with 1080p HD broadcasting, simple interfaces, and remotes.

The Chromecast is a lot cheaper, at a third of the price of Amazon Fire TV, but it also has approximately one third of the capabilities, making it an exceptional tool for smaller needs - like streaming from your laptop or tablet directly to the television. It is also a Google product, which means you will need to acquire a Google account if you do not already have one in order to use it.

Ultimately, Amazon Fire TV is the fastest streaming media device on the market today. It boasts a quad core processor and 4 times the memory of any other streaming media device. Because of this, it speeds through menus, boots quicker, and starts videos faster. Ultimately, this will save you a lot of time during your TV and video viewing over the life of the device.

You can use a Roku, Chromecast, Apple TV, and/or Amazon Fire TV to save money on monthly cable or satellite TV bills. My guidebook *How to Get Rid of Cable TV & Save Money: Watch Digital TV & Live Stream Online Media* provides additional tips and resources on ditching a monthly cable or satellite television service and using less expensive and sometimes more convenient alternatives to watch TV.

What is in the Box?

When you open the new Amazon Fire TV box, you are going to recognize everything inside as standard electronic devices and their necessary counterparts for operation. That is to say, nothing is going to jump out and spin in circles welcoming you to a new world of set-top entertainment, although there are a lot of things in store with the latest streaming device. Inside you will find:

- Amazon Fire TV Box
- Remote
- Power Cord
- Two AAA Batteries
- Quick Start Guide

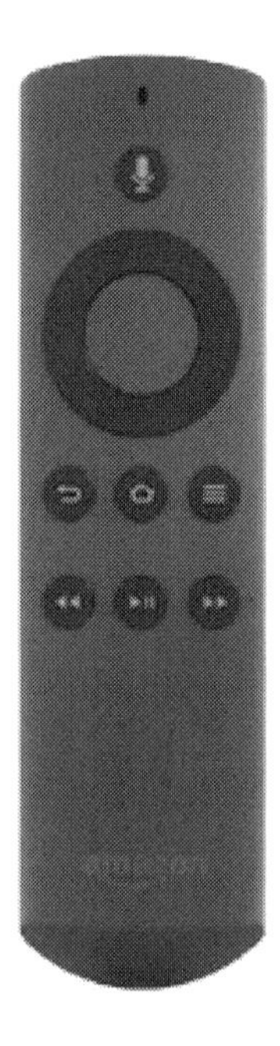

The batteries are for the remote, and everything else runs from the electricity in your home or office.

What Else Do You Need?

Before you can get started with the setup and the actual enjoyment of the Amazon Fire TV, you are going to need a few more things:

- HDTV
- HDMI Cable
- High Speed Internet

Most of these things already exist in modern homes, but just to be sure, check your television for its HDTV compatibility first. This will save you any trouble going forward, while allowing your enjoyment of your new television toy to start right away!

Initial Setup of Amazon Fire TV

The Amazon Fire TV arrives already pre-programmed to your Amazon account, so there is no need to enter any codes or alternative information when you begin connecting to your HDTV. The little box is ready to go, and all you have to do is connect it to the power source and television to get started.

Connect and Power Amazon Fire TV

There are only a few steps to setting up the Amazon Fire TV, and the process is amazingly simple.

1. Plug the power cord into the set-top box and the wall, so there is actual power running to it.
2. Plug the HDMI cord into the same box.
3. Plug the HDMI cord into an available HDMI port on the HDTV.
4. Make sure your television is on, and power up the Amazon Fire TV and allow it to walk you through the steps of connecting to your high speed internet.

Once you have connected to the internet, your menus will come into play and you will be able to navigate around everything that is so great about the Amazon Fire TV. Let's get started!

Fire TV Main Menu

From any screen on your device, you can press the "Home" button to return to the Main Menu. The Fire TV Main Menu consists of the following:

- **Search:** Here you can search for movies, TV shows, games, apps, and music videos from Vevo.
- **Home:** Allows you to review recent activity and content recommendations.
- **Movies:** Here you can rent, buy, and watch movies from the Amazon Instant Video store. Amazon Prime members can watch Prime Instant Video as part of the Amazon Prime Benefits.
- **TV:** Here you can buy and watch TV show episodes or seasons from the Amazon Instant Video store. Amazon Prime members can watch Prime Instant Video as part of the Amazon Prime Benefits.
- **Watch List:** Access your Amazon Instant Video Watchlist.
- **Video Library:** Includes all of the Amazon Instant Video movies and TV shows you've purchased or are currently renting. All content you have purchased is stored in the Amazon Cloud, and you can stream it to your Fire TV device. This library is only for Amazon purchased content.
- **Games:** Shop for, buy, and play games from the Amazon Appstore here.
- **Apps:** Here you can shop for and buy games and apps from the Amazon Appstore.
- **Photos:** Here you can access photos and personal videos from your Amazon Cloud Drive account. You can also start up photo slideshows and set photos as transitioning screen savers.
- **Settings:** Go here to manage your Amazon Fire TV apps, controllers, parental controls, Internet connection, and many more aspects of the device.

Fire TV Settings

The Amazon Fire TV comes with most configuration settings automatic, but you can use the "Settings" menu to manage your apps, Internet connection, controllers, screen savers, and more.

The following are on the Fire TV "Settings" menu:

- **Help:** Here you can access Amazon Fire TV Help videos along with quick tips, and customer service information.
- **My Account:** Here you can Register or Deregister Amazon Fire TV as well as Sync Amazon Content, so that it is available on your Fire TV device.
- **Second Screen:** Here you can turn on or off the "Second Screen" option for use with your nearby compatible tablets.
- **Applications:** Here you can access your Amazon GameCircle, Appstore, and installed application settings.
- **Parental Controls:** Using Parental Controls, you can block purchases and restrict access to movies, TV shows, games, apps, photos, and more.
  Note: *These Parental Controls do not apply to third-party applications. They only work for your Amazon content. Each third-party app may have its own parental controls available.*
- **Controllers:** Here you can pair or unpair your compatible Amazon Fire TV remotes and Bluetooth game controllers.
- **System:** Here you can view and manage device information and settings like screen savers, Wi-Fi network, display calibration, audio, and time zone settings.

Watch Amazon Fire TV

While Apple TV and Roku feature content "channels" on the streaming media devices, Amazon Fire TV features "apps" on the device. Some of these include Hulu Plus, YouTube, Netflix, Showtime Anytime, and Spotify.

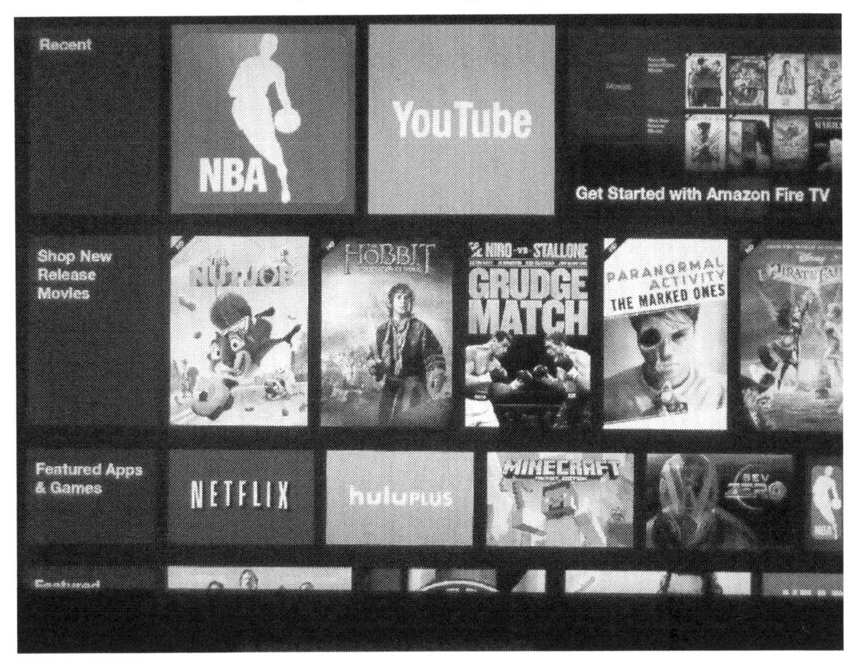

You can add apps from the device, based on your preferences. Use the following URL for all the currently available Fire TV apps
<http://www.amazon.com/b?ie=UTF8&node=7031433011>

To add apps on your Fire TV

To watch TV or play games using Fire TV, you will have to add the apps of your choice. Some apps are free while others are paid. It is entirely up to you which apps you decide to add. Complete the following steps to add apps.

1. Using your remote navigate down the left side main menu and choose "Apps."
2. There are nine categories you can add apps from as of this publication:
   - Cooking
   - Entertainment
   - Finance
   - Games
   - Health & Fitness
   - Music
   - News & Magazines
   - Photography
   - Sports
1. Each of these categories contains a number of unique apps, many of which are free, and some of which may cost you to purchase with your Amazon account.
3. Once you're in a category and have found an app you'd like to add, choose that app with your remote. You'll see the app's information screen which will give you a description of the app, several screenshots, and info about whether it is free or costs money to buy.
4. On the price area, where it may say "Free" or show the "price," choose this button to install the app to your Fire TV. It will now say "Processing" and then "Downloading" as it works to add the app to your device.

Once the app has fully downloaded, you will receive an "Open" option on the app screen and you can open the app on your Fire TV to use it.

Apps will appear in your "Apps Library" on the Main Menu. You can access this from the Main Menu whenever you go to the "Apps" option on the left side of the screen.

Using the Remote

The Amazon Fire TV comes with a remote in the box complete with the batteries it requires.

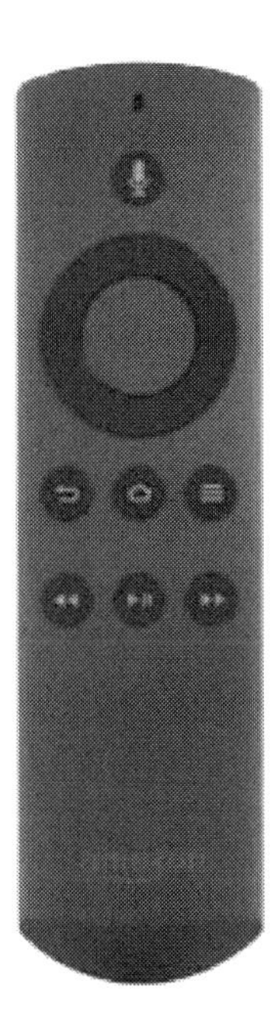

The following information describes each button on the remote, which is shown in the previous image, and what it does.

Top microphone icon - allows for voice search for content by title, or people's names. You can search for movies, actors, TV shows, songs and artists, as well as apps you've installed.

Smooth circular button - allows for navigation to move up/ down and right/ left on various screens of Fire TV. Inside the button is an "OK" button that can be pressed to make selections on the various screens.

There are six small buttons towards the middle of remote. The following are descriptions of the buttons from top left to bottom right on the remote.

Go back - Used to return to a previous screen on your device.

Home - Used to return to the main home screen of your Fire TV.

Options - Used to bring up options menus on various screens and apps.

Back/rewind - Used to rewind media content back.

Play/Pause - Used to play or pause media content.

Forward - Used to advance forward on media content.

Renting or Buying Media

Whether you decide to rent a movie from Amazon.com or purchase it outright to become part of your library, it will automatically be stored in your Video Library for viewing.

Prices vary, based on the selling or renting price per movie or episode of your favorite television show, but each purchase will remain in your Video Library for the duration of its validity. That means if you buy it, it is yours. If you rent it, you will be allotted a certain amount of time to watch it - usually in an unlimited capacity - before the rental period expires.

Music, Movies, and TV shows

Think of your Amazon Fire TV this way: If you can purchase, rent, watch, or listen to anything available on Amazon.com, you can do the same with your new device. Music, movies, televisions shows, and even your own personal videos are available on the device, and can be accessed effortlessly at any time.

You can even hide channels from your kids - should they be so inclined to snoop through prohibited or sensitive material - and set up restrictions for anyone but you to operate the device. All you have to do is password protect purchases and channel access through the device by going to "Settings" and accessing the "Security" segment of its operation. The intuitive design allows you to work the screen just like you would a tablet or smartphone, so you never feel lost when making changes or decisions along the way.

To access the Parental Controls, simply start at the Main Menu of your device, scroll to Parental Controls and update the settings to restrict access to content and purchases from Amazon. However, currently, third party purchases cannot be blocked – like those from Hulu and Netflix. If you do not currently have a PIN number for Amazon purchases, you will need to create one at Amazon.com/PIN to use the Parental Controls accurately.

Setting up Restrictions for Purchases and Rentals

The Fire TV allows you to set up restrictions for apps, in app purchases, Amazon Instant Video, and to hide or block content. The way to restrict content and purchasing is by setting a Parental PIN.

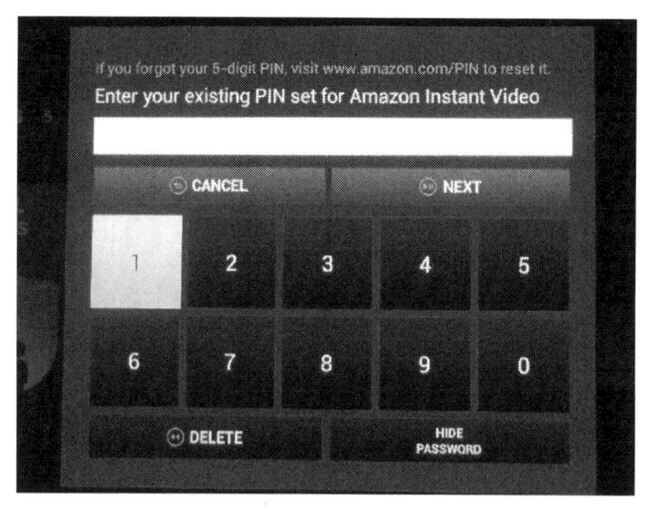

To set a Parental PIN on Amazon Fire TV, complete the following steps:

1. On Home screen, choose Settings, then Parental Controls.
2. Press the Select button again ensuring the button is set to "ON."
3. Enter your current Amazon Parental Controls PIN (if you already have one) or create a new one at www.amazon.com/PIN, and then select "Next." The following Amazon Fire TV parental control options will display:
   - Require a PIN for all purchases.
   - Require a PIN for Amazon Instant Video only.
   - Block the ability to view or purchase certain content types, such as games, apps, or photos.
   - Change your Parental Controls PIN.
2.
3. **Note:** If you somehow forget your Parental Controls PIN, you can reset it by visiting www.amazon.com/PIN.

Amazon FreeTime for Fire TV Coming Soon

Amazon FreeTime already works on your Kindle - if you have one - by giving your children, or even other family members - limited access to what goes on within the electronic device. This means apps, channels and content will all be personalized with your login information. You will also be able to set up daily screen limits and restrict certain types of content from streaming altogether, giving you the power over the device's full capabilities.

The good news is, if you do have little ones running around, there is FreeTime Unlimited on the way that caters to children ages three to eight years old, delivering their favorite movies, television programming and more for just $2.99 per month.

This feature is expected to become available for Amazon Fire TV in May 2014.

Amazon Fire TV Apps Overview

The Amazon Fire TV does not have channels. Instead, it has apps where different types of content are located. The following is a description of several of the different apps currently available on the Fire TV. As time goes by, there will most likely be many more apps available for the device from Amazon and third parties.

The revolutionary Amazon Fire TV is able to deliver both local and international content for its users. This is a wonderfully designed streaming box that garners attention from everyone due to its incredible platform and exciting possibilities for future apps.

At its release, the Fire TV has a wide range of quality applications that include the biggest names in the business, and it will continue to transcend to greater heights. This is a system that is on its way up, and there is no doubt Amazon will continue to update and upgrade the Fire TV applications.

Check out a description of the TV, video, and movie apps for the Fire TV in the next sections.

Amazon Prime Instant Video

Amazon Prime, or Prime Instant Video, is a media streaming service that is brought to you by the ever popular online giant Amazon.com. This exciting service is available for $99 per year, and includes unlimited access to movies and television shows. The benefit of Amazon Prime is that its membership price comes with a free two-day shipping upgrade on any item you order from Amazon, and access to thousands of eBooks.

The downfall is that its media streaming services include popular movies, but not always the newest releases. You can dip into the archives of some of your favorites, but if you are looking for the movies that are new to DVD each Tuesday, this service lacks that up to date availability. It does, however, bring you past television episodes and kid's programming at no extra charge, but will not provide you with the previous day's television programming for free. You can watch all of their offerings as often as you like, even if you want to watch the same episode from the first season of *Downton Abbey* for the thirteenth time. Just do not expect to see the current episodes for quite some time.

Netflix

Netflix is available to subscribers, and your account must be up to date to enjoy its content. For $7.99 a month, you can become a member and enjoy the seemingly endless supply of content that is available on demand, whenever you are ready to watch! With your Netflix subscription you will receive complete access to all of your favorite on demand selections, and literally thousands of movies and television shows for your enjoyment.

With Amazon Fire TV you can stream movies, television shows, trailers and original programs directly to your larger screen instead of using your tablet or computer as an entertainment source. With thousands of streaming options available on Netflix, there is no end to the content you can enjoy directly from your Amazon Fire TV box.

Hulu Plus

For the same price as Netflix, $7.99, Hulu Plus users can enjoy their favorite TV shows on demand, and even new network offerings as early as the day after they aired originally. The Hulu database includes over 2000 shows, movies, trailers for new and upcoming movies in the theater, and promos for your favorite shows so you know what to watch for! Check out the Hulu website at www.hulu.com for more information.

Showtime Anytime

Customers who have a subscription to the Showtime lineup of channels on their cable or satellite TV package can enjoy the use of Showtime Anytime on the Amazon Fire TV. This service features hit movies, sports and other programs as well as the ability to watch whatever content happens to be airing live on Showtime at the moment.

Other cool features of the service include "My List" to organize or view your favorite programs, "Add New Episodes," so you'll get automatic updates of any new updates for your favorite shows and "Play Shifting" to let you resume watching a program from where you stopped on another device. Keep in mind that use of this app on the Fire TV will only work for customers who have Showtime through participating TV providers.

YouTube

There is nothing better than navigating the hilarity - and the actual informative content - that is available on YouTube. Instead of playing the website's content on your device, you can literally stream it through Fire TV, giving you all of your favorite videos directly on the big screen. Also, if you have a YouTube.com account, you can access all of your favorites right away simply by logging in.

Crackle

Crackle is a Sony owned company that provides original web content, feature films, and television programming that is produced by Sony Pictures. Much like Hulu Plus or Netflix, users are able to enjoy popular films that exist within the Crackle catalog. The catalog updates monthly, with new content being added and old material being removed to keep the site running tightly.

Crackle provides a different take on movies and television series, providing full length, uncut and unfiltered content to the masses. You can choose to watch a block of Seinfeld episodes, or watch an old Bruce Lee movie at the drop of a hat.

Crackle also provides original programming options, but no matter what you choose it is always free. Simply download the app to Fire TV, register, and set up a profile to get started. You can create a watchlist, restrict access to minors, and even share your findings, episodes, or movies with friends and family members.

WatchESPN

WatchESPN is a great outlet for sports - but only for cable subscribers who already receive ESPN as part of their programming package. If you do, you are in luck and are able to watch all ESPN3 content from your Fire TV, as well as enjoy simulcasts that are currently playing on ESPN, ESPN2, ESPNU, ESPNEWS, ESPN Deportes, ESPN Goal line and ESPN Buzzer Beater, so you will never miss out on your favorite sporting event again.

NBA GameTime

This is the official application of the NBA. It is an award-winning application that has earned rave reviews for what it has to offer both the casual and hardcore fan. With a collection of statistics and related data, it just does not get any more in-depth than this application.

For users who want to get the latest statistical and visual updates, the amount of information that is accessible cannot be overlooked. For those who want to follow the press conferences after games live, they are able to do so with a tap of a button. There are also various highlight videos available on demand including top plays, game recaps and more.

Following your favorite NBA team has never been easier than it is with this app. Users are able to access a range of highlights and information on the go without having to venture away from their TV.

NBA GameTime brings the game to the fan in a modern, interactive manner. This app can also connect to a paid NBA League Pass account to watch live streaming NBA games during the season.

Thuuz Sports

This is a comprehensive sports application that keeps fans updated on their favorite sports teams. The application rates each game from 1-100 to signify how important it is in the grand scheme of things. It also updates stats and ensures users understand the importance of each and every game.

Are you thinking about catching the action on a nearby TV, but are unsure where to look? Each game has details like the broadcasters and what channel the game will be, or is currently appearing on.

For those who just want alerts from their games, the highlights will come in the form of notifications as soon as they happen in game.

RedbullTV

RedbullTV provides access to sports, music, and live entertainment for free. The channel focuses on extreme sports, and other twists and turns that may not qualify under the same genre, but are nonetheless exciting to watch (including a winch slip and slide "ride"). Surfing, BMX, motorbikes, and snowboarding are all available in extreme capacities with this app, as well as some insightful original programming from promoters, athletes, and professionals in the sports and music industries.

DailyBurn

Are you thinking about getting up and working out, but are unsure about where to begin? DailyBurn is home to a range of workout programs designed to cater to the needs of the user.

Whether you want to lose weight or gain muscle mass, there are workout programs designed to meet those and other needs. There are over 100 full-length workouts that are accessible within moments.

All of these workouts come in the form of videos to engage you and make it easy for you to follow the exercises. This is a comprehensive application that is designed by professionals for excellent short and long-term results.

Bloomberg

Bloomberg television is a 24-hour-a-day financial news outlet that provides up-to-date information affecting the business and financial world, including trading, tickers and informative broadcast analysis. This app is available by cable or satellite subscription only, and can be viewed from your Amazon Fire TV once your account information has been entered.

Huffpost Live

In the modern age of news, it has become imperative to get up-to-the-minute coverage on all the main events around the world. Whether it is politics, business, or sports, the Huffington Post is a top source for the latest in what is going on in the world.

This app features experienced correspondents who are ready to provide updates within minutes as well as ongoing coverage.

Users are able to comment on articles, and have their opinion heard on the spot. It is an interactive, updated app that is ready to provide great coverage from all parts of the world.

Vimeo

Sharing and storing videos has never been easier than it is with Vimeo. This all-encompassing app is designed to ensure all your videos are integrated at the highest quality.

All videos can be played through the application after they have been updated. This is instantaneous and flawless for those who seek perfection from their player.

Users do not have to be online to watch their videos and they can go through them while they are offline too. It is easy to swap through the app to watch and share videos because of its smooth, effortless interface.

Do you want to share videos with family and friends in mere moments? It is easy to do this with Vimeo due to its 'AirDrop' functionality.

This is an easy-to-use application that ensures playback for videos has never been simpler. A well-designed, simple application that should prove to be worthy for those who have a large collection of videos.

RealPlayer Cloud

RealPlayer Cloud allows you to move, watch, and share your videos among all your devices. No longer do you have to worry about formatting or converting videos, or carrying HDMI and USB cables in order to watch your own videos or share them with others.

The great news is when you sign up, you get 2 GB of space for free. See the Real.com website for more details on the service and its integration with your Fire TV.

Flixster

Are you a diehard movie nut? Are you craving having movie trailers in easy reach? Flixster is home to an updated assortment of information and trailers to keep all movie fans in touch with the latest films.

Do you want to watch what that upcoming movie is all about? Flixster has HD quality trailers that can be watched to assess how good the movie might be.

Are you thinking about watching a specific movie and want to know its show times? The app has all local show times updated to its program and can be accessed at a moment's notice.

Flixster is an incredible, in-depth app that is designed to make Hollywood come to life in the comfort of your own home. With critics' reviews, quality trailers, and even full-length movies, Flixster is a must-have application for die hard movie fans.

Plex

Do you have a large collection of personal media that needs organizing? Plex is the number one source for all audio and visual personal media that has to be put together. Users can collate all necessary personal media files and organize them based on personalized titles, artwork, and much more. The options are endless with Plex when it comes to putting together a file of media that is easy to access.

Are you thinking about playing this organized personal media on the go? With Plex, users can easily access their personal media and watch/listen to it on a host of different devices with ease.

Plex enables users to share these files with family and friends through the tap of a button. It is a convenient solution that is designed to make the world come to life and connect everyone through their devices. Users are also able to access a range of online channels such as TED and TWiT for even more content. The content is always being updated, giving plenty of great media content for viewers to peruse.

AOL On

Watching web videos might have been revolutionized by YouTube, but it has been kicked up a notch with AOL On. This application is the go-to solution for users wanting to search for and watch quality web videos.

With trending topics, breaking news, and entertainment videos, AOL On is one of the best applications on the market for premium content. The application is home to 14 diverse channels ranging from technology to style. There is something for everyone.

Sharing videos is easy with AOL On, which makes the entire experience interactive and unique. Family and friends will also be able to watch those exciting videos from the comfort of their devices.

TastyTrade

Are you trying to stay in sync with the modern financial world? TastyTrade is the number one application available to keep users updated to the rising and falling of stocks around the world. With heaps of financial data and graphs galore, TastyTrade is the finest source of information available for enthusiasts.

Real traders are on hand to explore and discuss new strategies of how to maneuver around the financial market and see gains. This is an exceptional application for those who want to see the financial world unfold in front of their eyes from home using their Fire TV.

Smithsonian Channel

Just as you would imagine, the Smithsonian Channel's content is inspired by their research and museums, and it gives viewers' full access to clips, full episodes, and content that is available on an array of their networks.

The subject matter of the content on this app includes original non-fiction programming that covers a wide range of historical, scientific, and cultural subjects. It is only available to viewers who currently subscribe to the channel via cable or satellite enrollment, so you must validate your service using your login information.

Classic TV

Are you feeling nostalgic? Do you want to go back into the 60s, 70s, or 80s? The options are endless with this Fire TV application.

Classic TV offers some of the finest TV out there, and it can help rekindle fond memories from your past. Watching modern TV is not what everyone wants to do in their spare time. Being able to go back and watch some of those classics can be a scintillating experience.

With this app and its wide range of classic TV shows, the choices are simply endless. The database is constantly being updated to meet the visual needs and wants of users.

iFood.TV

This application is a foodie's dream come true. With an array of recipes, tips, and relative information, the world of delicious food has never been more accessible.

Users can immediately hop on and sift through a wide range of recipes to have it pop up with a tap of the button. With over 40,000 recipes, the list of menu options provides nearly a lifetime's worth of eating. The recipes come in all types and are continually updated to make sure the user has something new to cook every meal.

Get into the kitchen and start cooking along because the recipes will not stop coming. The perfect meal is out there and this application is ready to point the user towards it. When it comes to excellence and food, this Fire TV application has it all.

Tubi TV

Are you craving a good movie or TV show? Tubi TV is the number one option for those who want to watch the latest movies and TV shows from the comfort of their own sofa. There are thousands of titles to sift through, and the list is continually being updated to meet the needs and wants of those movie enthusiasts.

The quality of this app is impressive, the movie lists are extensive, and the visuals are delightful. It is a movie enthusiast's dream come true. Tubi TV brings the world of films to your Fire TV in a moment's notice. Users will be left drooling at all of the high-quality, renowned options available to them via this app.

Redux TV

Redux TV is home to an organized, modernistic approach to watching television. It encompasses all that is right with traditional cable and puts a twist on it.

Users are able to sift through curated lists that have been put together by professionals based on a range of specifications. Each genre has its own collection of items that are easy to navigate and find through the application.

This makes watching TV much easier and simpler than ever before. Just find the right list and watch away as the choices are simply endless when the list is selected.

All Fitness TV

Joining a gym is not necessary in the modern age, and this application is testament to this fact. All Fitness TV is an all-encompassing app that pinpoints the intricacies of workout programs into one extensive package.

Users are able to quickly sift through the collection of workout programs and pick the one that best suits their specific needs. There is an extensive library of videos geared to teach and guide users towards a healthier and fitter lifestyle.

This application takes fitness to an entirely new level all from your own living room. It just does not get better than this!

Quello Concerts

Quello provides a focused approach to entertainment by streaming only concerts and music documentaries. This app licenses a variety of long-form concerts, documentaries, behind the scenes footage, and interviews from both major and independent music labels. The musical genres are versatile and plenty, ranging from the 1920s to today, with each category covered with in depth original programming.

NowThis News

Are you not a fan of those local news channels that keep repeating the same information again and again monotonously? NowThis News is the number one option for users who want immediate, up-to-date breaking news coverage that is unbiased and free.

Whether it is breaking news, sports, business, or politics, this application has it all and then some. The range of topics are endless and the options are continually growing. The videos are original and to the point just like all users want them to be. There are approximately 50+ new videos released everyday on a range of topics to keep the user updated.

This application not only relays the information that is going around the world, but also breaks it down for the user. This adds context to the news that is being presented and creates intrigue as to what is actually taking place and the effects it will have around the world.

Hasbro Studios

Hasbro Studios proves original family and children's programming through their app channel. If you grew up in the 70s, 80's or the 90's, then you know how important and iconic the Hasbro brand is. Hasbro is the toy manufacturer and cartoon creator that brought us GI Joe, My Little Pony, Jem, Pound Puppies, and so many other great TV shows that were not only entertaining, but also delivered a great message about family, friends, honesty, and being a good person. These are the type of shows that never grow old. Each generation of children will find something appreciate here.

FitYou

One thing that you will notice on TV is that there aren't as many fitness shows as there once were. In the late 90's, there was at least two hours of fitness programming available on cable TV, but as of 2014 there isn't a single daily running fitness show that is broadcast in every market.

With FitYou, the Amazon Fire TV provides you with an app that offers fitness programming 24/7. When you are ready for a workout, this app will have just what you need. Browse through its multiple selections to find the best program for your fitness needs.

Happy Kids

The Happy Kids TV app is familiar to anyone who has children. It is to this generation what PBS children's programming was to past generations. Happy Kids TV provides some of the best educational TV for children. The app channel offers entertainment and educational programming, as well as workbooks and assignments that enrich the viewing experience.

This is an app that you can feel comfortable allowing your child to watch on their own and it is the type of television that is actually good for kids. The shows have lessons on friendship, responsibility, math, science, music, and entertainment. If ever a channel was a digital teacher, the Happy Kids channel fits that role.

You may have noticed that beyond PBS there is very little educational TV for children. Sure you can turn to Nick or Disney, but most of those shows are cartoons, kiddie sitcoms, and drama. The Happy Kids app is one to watch for anyone who has children and wants to feed their minds with educational content while also entertaining them.

Post TV

Do you miss real journalism and has the 24-hour news cycle become more about "infotainment" than what truly matters? While newspapers around the globe are struggling, the Washington Post has found a niche in hardcore classic journalism. Post TV is an extension of what goes on in their newsroom around the clock.

On this app they discuss everything that is important in Washington, and they explain why. Few news programs explain why certain things are important, but Post TV does just that. This is an app that any news junkie will love, and it is a great departure from the CNN's, MSNBC's, and Fox News-like programming in the world. It is a raw news channel that asks the big questions and that digs deeper than the average cable news channel dares to dig.

Hollywood Ticker

The Hollywood Ticker app is the Amazon Fire TV answer to entertainment channels like E! The Hollywood Ticker is a 24/7 entertainment gossip channel where you can find out all the latest dealings in Hollywood.

If you care about knowing what the rich and famous are up to, then this is your channel. In many ways, it is like E! before it became centered around reality shows like "Keeping Up With The Kardashian's." It is more Jewel Asner and Brooke Burke era E! vs. Ryan Seacrest era E! This is a great app for raw entertainment news.

MBC On Demand

The MBC on Demand app brings viewers a channel that offers subtitles to the best Korean TV shows, dramas, soaps, and movies. If you are a fan of Korean shows, MBC On Demand is perfect for you and your friends. It will provide you with all the Korean drama you can handle!

ACC Digital Network

College sports fans will certainly love this app. When it comes to sports, there really isn't an off season for the media or the fans. If you are like most hardcore sports fans, you want to keep up with everything that is going on with your favorite teams or conference.

The ACC Digital Network gives you access to every team and every sport that is played in the Atlantic Coastal Conference (ACC). ACC sports from hoops to field hockey are covered on this network. It is a true fan's fantasy come true. This app is for the people who want to learn more about the ACC players, the schools, the coaching staff, and what they are working on. It gives greater insight into how college athletics runs.

If you like ESPN, then you will love this app and what it has to offer an ACC die-hard fan.

Music on Amazon Fire TV

Music comes to life on Amazon Fire TV and delivers access from all of your favorite places, including the Amazon Cloud, Pandora, iHeartRadio, TuneIn, and videos from Vevo.

Amazon Cloud

The Amazon Cloud functionality on the TV allows you to access all digital music - songs and albums - you have previously purchased from Amazon.com as MP3s. You can even enjoy your musical selections while scrolling through images or playing games, so there is no end to the perfect soundtrack to your day.

If you are currently an iTunes person and Amazon Cloud has never been the focus of your musical library, there are a number of other musical outlets available on the Fire TV, so you do not have to worry about crossing platforms.

Pandora

Pandora Radio is a music app that allows listeners to search for musical selections by entering song titles or artists they love, and receiving a radio "Station" as a result.

If the user enters "Brad Paisley" as a search artist, it will also play artists like Toby Keith and Kenny Chesney, and provide those musical selections as a "Station." So if someone says they are listening to Kenny Chesney radio on Pandora, it is simply the result of entering his name, and receiving music that relates to the genre and sound. Pandora also provides a feedback system that allows users to give positive and negative evaluations of the musical offerings, to hone their specific listening experience.

Listeners can purchase songs or albums too, but absolutely do not have to in order to enjoy the service. There are two subscription options available with the Pandora app. First there is free service, which is supplied through advertisements, and second there is a no-ad fee-based subscription. The Pandora library is comprised of 800,000 tracks, from over 80,000 artists.

Vevo

Vevo goes a step further than offering radio station-style music, as it literally brings your television to life with full length music videos! Watch your favorite music videos and discover new ones on the VEVO app for Amazon Fire TV. This app provides access to VEVO's entire catalog of 75,000 music videos from more than 21,000 artists.

Who needs MTV or VH1 when you have VEVO at your fingertips? The best part is you are not subjected to videos in the order the station wants to play them. You can physically pick and choose your favorites to watch over and over again, or you can display new versions you haven't seen yet. You can even invite your friends over to share in the mystery of what was once one of the best ways to enjoy music: Through music videos!

Tunein Radio

TuneIn Radio provides both a local and international approach to listening to music, sports, news and entertainment offerings from around the world. You can choose from live global newscasts to international radio stations and specialty podcasts that you cannot get anywhere else. Listen to your favorite local stations digitally or look for radio stations to listen to by musical genres. This is a great free app to enjoy music from your local regions and around the world!

iHeartRadio

This Clear Channel owned radio station provides aggregated radio content from over 800 different radio stations, making the culmination of content available to users. The content is completely free to enjoy, making for a great listening experience on your Amazon Fire TV with this music app.

Gaming on Amazon Fire TV

The Amazon Fire TV also offers gaming, which can be fully unleashed with the additional Fire TV Game Controller. Along with the controller is a wide selection of fun and exciting games available for play, making your Fire TV into an interactive gaming console!

Some of the games that are available with just a few clicks include the following favorites:

- Minecraft
- The Walking Dead
- Monsters University

Getting Other Games

With over one hundred games available on the Apple Fire TV, the average cost of paid games is only $1.85, meaning the affordability factor is real and already in your living room. Some games are even free, which makes it even better! Games can be downloaded as effortlessly as movies and music, and are playable immediately upon their download.

You will see several familiar titles for the Fire TV, and the good news is they cost significantly less than they do on platforms like Sony's PS4 or Microsoft's Xbox One.

The following are some of the game apps currently available for the Fire TV:

Despicable Me: Minion Rush: Free

In this app, Minions from the Despicable Me franchise will Run, Jump and Dodge obstacles through the Downtown area, which features favorite places from the movies. Also, there is a brand new secret area called The Bank of Evil! In this fun Fire TV app, you can customize your Minion with unique costumes, weapons, and power-ups.

Sonic CD: $2.99

In Sonic CD, everyone's favorite hedgehog travels to the distant shores of Never Lake to see the annual appearance of Little Planet. Dr. Eggman tries to steal the Time Stones while Sonic battles to stop his nemesis from succeeding in his evil plan.

This app is just one of several Sonic the Hedgehog apps available for Amazon Fire TV gaming.

The Walking Dead: The Complete First Season: Free

This is the first episode of a five-part game, and episodes two through five are available via in-app purchases. You will play the game as convicted criminal Lee Everett, who has a second chance at life. Your choices in the game will affect how the story ultimately plays out.

Asphalt 8: Airborne: Free

Choose from 47 luxury cars in this extreme fun racing game. Race in exotic venues such as Venice, Guiana, and the Nevada Desert. In all, there are nine from which to choose. This game also features simultaneous multiplayer action, or you can choose to challenge your friends to asynchronous races.

Minecraft - Pocket Edition: $6.99

You will place blocks, build things, and go on adventures in this pocket edition Minecraft app. It supports 5 Player Local Network VS as well as Co-op mode. Choose from either Survival or Creative modes, and craft, create, and breed anywhere.

Zen Pinball HD: Free

This app features a huge selection of pinball tables, which include the critically acclaimed Marvel Pinball series. Additional tables are available via in-app purchases. You can play multiplayer games 4 Player Local Pass. This app is a must have for pinball lovers.

The Game of Life: $4.99

Based on the classic board game of the same name, this app allows you to Spin the wheel of fate and live it up. The graphics, visual effects, and sounds are all superb in this app, which features easy-to-use controls and clear play instructions. Playing the game with players Mary, John, Rachel, or Bob, anything can happen in The Game of Life.

Monsters University: $0.99

This app features two exciting games with characters Mike, Sulley, and Squishy from the Monsters movie franchise. As you level up through the game, you will unlock exciting power-ups to improve your game performance. Fans of Monsters will be big fans of this exciting app.

NBA 2K14: $7.99

This is the latest installment in the NBA video game franchise. With this app you can play through multiple NBA seasons while establishing your team as a dynasty. This app features the full NBA roster.

Crazy Taxi: $4.99

Featuring original music from The Offspring and Bad Religion, Crazy Taxi takes users on a wild ride through traffic-packed streets. You can play this throwback to 2000 app in either Arcade Mode or Original Mode. The craziest taxi driver wins.

Shadow Fight 2: Free

Combining classical Fighting and RPG, this app provides the ultimate gaming experience. Throughout the game, you will battle foes and demon bosses in order to close the Gate of Shadows. You can customize your fighter with epic swords, nunchacku, armor suits, and magical powers.

Tetris: $4.99

Tetris is a highly addictive puzzle game. This Fire TV app offers Marathon Game Mode, which consists of 15 levels. Combine distinctive geometric shapes to eliminate rows of blocks before they reach the top of the screen.

Fire TV Game Controller Setup

The additional Fire TV Game Controller can be purchased as an accessory to enhance your enjoyment of the Amazon Fire TV. The controller uses Bluetooth and easily pairs with your Amazon Fire TV. Also, if you have a Kindle Fire HDX tablet, it can be paired with that as well. The controller includes a set of instructions in the box about how to do this, along with two "AA" batteries. Here's a look at how to pair the controller with your Fire TV and its functionality.

To pair with the Amazon Fire TV:

1. Make sure two "AA" batteries are inserted correctly into controller.

2. Hold down the "Home" button on the game controller for about 5 seconds. Make sure you see the status lights on the front of the controller blinking. They'll blink two on one side and then two on the other side, repeatedly.
3. With your Fire TV turned on, go to the "Settings" option on your main screen.
4. Choose "Controllers" and then "Bluetooth Gamepads."
5. Select the "Amazon Fire Game Controller" if it isn't already selected.

You should now see that your controller is paired and controls your Fire TV.

Note: *You can also unpair the game controller in your Fire TV "Settings" and "Controllers" by going to "Bluetooth Gamepads" and pressing the button on your game controller that has 3 vertical lines.*

Using the Game Controller

The Game controller has a variety of buttons on it and almost resembles the layout of a PlayStation 3 or PlayStation 4 game controller. The controller not only helps in terms of playing games, but can also be used as a remote for your Fire TV.

On the front of the controller, which is pictured in the image above, there are R1 and R2 as well as L1 and L2 buttons. In some games these are called "trigger buttons." For example in the game "Riptide G2," these will help accelerate and decelerate your vehicle.

On the controller surface to the very left is the Left Thumbstick. This is generally used to move characters or steer in games. There is also a directional pad right below the left thumbstick. This is also used for moving around in games or navigating menus.

The middle of the controller has similar buttons to your Fire TV remote. The first is the "Return" to previous screen. Next to that is your "Home" button, and finally there is the "Options" button.

Beneath those three buttons is the "GameCircle" button. GameCircle is a way to track your achievements on games and in some cases a way to synchronize your latest progress on a game so it isn't lost.

On the right side of your controller are the Y-X-A-B buttons used in varying combinations for playing games. Each game may have different uses for these buttons. Below those buttons is the right thumbstick, which is also used for steering and movement in many games.

Finally, at the very bottom of your game controller are reverse, play/pause, and forward buttons, which are used for reversing through, playing, pausing, and advancing through content on your Fire TV.

By pressing the GameCenter button during certain games, you'll bring up the GameCenter on your screen showing you various stats about game play and achievements. These may include the amount of time you've spent on the game, any achievements you've earned, and best high scores. In addition, there are leaderboards reflecting the high scores achieved by you and others on the game.

Tip: *If you press the "GameCenter" button on the Fire TV game controller on the main menu, it immediately takes you to the games selection on the left side of the Main Menu.*

Amazon Fire TV Tips and Tricks

All electronic devices have tricks to help you use the technology with ease and fluidity, and the Amazon Fire TV is not any different. There are a number of helpful tips that will allow you to enjoy the device and all of its capabilities, without pulling your hair out! The following sections describe a few of the best tips and tricks for this device.

How to Use Voice Search

Voice search is a great feature that allows you to use your remote to search for content using your voice, instead of navigating the system manually. All you have to do is hold in the microphone button and tell the remote what you are looking for, whether it is a television series, movie, app, game, song or musician you are searching for and appropriate results will automatically be displayed for you.

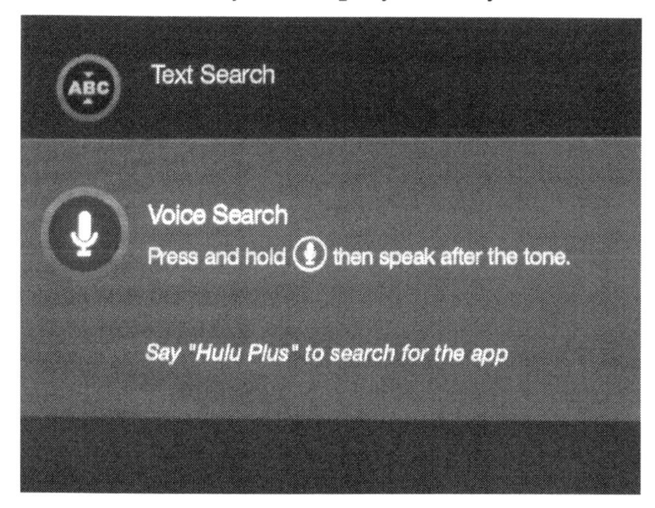

The Amazon Fire TV voice search works amazingly well. To use voice search do the following:

1. Press the microphone button on top of remote. A prompt will show up on your TV screen instructing you to hold down the microphone button and speak after you hear a tone sound on your TV.
2. Speak the name of an actor, movie, TV show or other type of content. The Fire TV will come back with what it thinks you said.
3. Press the "OK" button on your remote to confirm. Amazon Fire TV will bring up a directory of movies and TV based on what you spoke.

How to Use Amazon FreeTime

Amazon FreeTime for the Amazon Fire TV will be available in May 2014. This digital book will include a full update with exactly how to use Amazon FreeTime as soon as the service is available on Amazon Fire TV.

How to use ASAP

ASAP, or Advanced Streaming Prediction, is an exclusive Amazon Fire TV feature that predicts your viewing habits and automatically buffers the programs it believes you will want to view, so there is no delay in watching the television shows, movies, or videos when you click on them. Of course, the more you use Fire TV, the more it will know about your viewing habits, allowing this feature to work with precision.

How to use Second Screen

Second Screen allows you to view the content on your tablet - including video, audio, images, and applicable content - on the television screen your Amazon Fire TV is hooked up to.

The Fire TV works seamlessly with the Kindle Fire HDX to either mirror your tablet's screen, or allow you to use the KFHDX Second Screen abilities. Using this, you can easily share videos, movies, pictures, and more straight from your tablet to everyone in the room. Another great feature is that with Second Screen, you will be able to take advantage of the enhanced X-Ray feature.

To activate either of these features complete the following steps:

1. On your Fire TV settings, make sure "Second Screen" is turned "ON."
2. On your Kindle Fire HDX tablet go to "Settings," then tap on "Display & Sounds."
3. Tap on "Display Mirroring" and your tablet will search for any nearby compatible devices, such as the Fire TV.

When the Fire TV shows up on your tablet as an option, tap on the device name or "Connect." It could take about 10 to 20 seconds for the connection to be made. Your tablet's screen will now show up on your Fire TV, allowing for a truly interactive experience.

Note: To stop mirroring, simply press any button on your Amazon Fire TV remote. You can also swipe down from the top of your tablet to reveal "Display Mirroring" and you can tap "Stop Mirroring" to end the connection.

How to Mirror Tablet

If you are using a Kindle Fire HDX, you can mirror what is happening on your tablet onto your television screen - even if all you are doing is searching the web, or watching YouTube videos with a group of friends. Mirroring your tablet will require a few accessories:

- Kindle Fire HDX
- Miracast-enabled Accessory or TV (or the following)
- HDTV with HDMI Port (if you don't have a TV with Miracast)
- HDMI to HDMI Cable
- Active Wi-Fi

Plug the HDX into your television using the appropriate cable, or using the Miracast option. Once you have, locate the Settings button on your device, locate Displays & Sounds and tap "Display Mirroring." If you not turned your Wi-Fi on, you will be notified automatically. Otherwise, the mirroring will begin.

How to Set Up and Use Bluetooth Devices

Sometimes you need an accessory to make your television viewing life complete, and with the Amazon Fire TV you can pair a remote with your device. In addition, you can pair the gaming controller, which is available on Amazon for $39.99 and allows you to turn your streaming media device into a true gaming console as well.

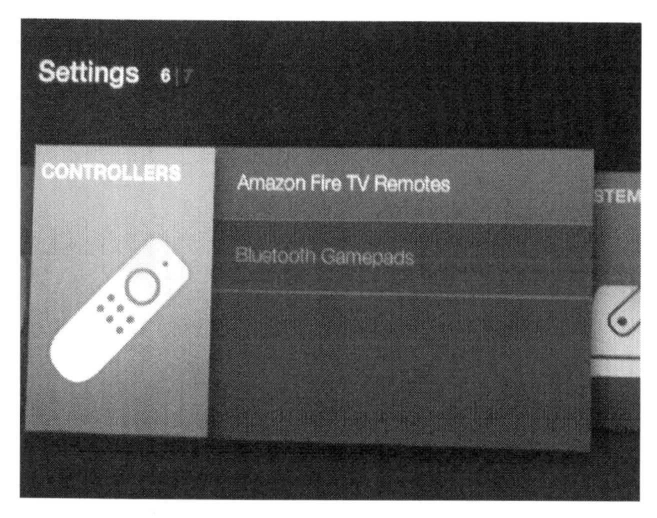

To add a Bluetooth device to your viewing pleasure simply:

1. Begin at the Home Screen of your device, and locate the "Controllers" segment by scrolling left on the screen.
2. Tap Bluetooth Gamepads.
3. Follow the directions onscreen to connect the device in question (keyboard, controllers, etc.).

How to Remove Content from Fire TV

It is possible that you have content you no longer use, but it still appears on your Amazon Fire TV. The good news is, you can remove these items you no longer use. Any purchases made from Amazon are saved to the Cloud and can be downloaded again to your Amazon Fire TV.

Note: *Individual app settings or in-app items may be lost if you choose to remove them from your Amazon Fire TV.*

There are three areas from which you can remove content on your Amazon Fire TV.

To remove content from your Watchlist simply

1. From Watchlist, go to a movie or TV show.
2. Select Remove from Watchlist.

To remove content from your Recently Watched sections simply

1. From Movies or TV, go to a movie or TV show.
2. Select Remove from Recently Watched.

To remove content from your Recent Carousel simply

1. From the Home screen, go to the item you would like to remove.
2. Select Remove from Recent.

Note: *Featured or Top content on the Home screen cannot be removed from Fire TV.*

How to Update System

Keeping your Amazon Fire TV up to date with the latest software will be key in its optimization and overall use. The device will notify you of any updates, but should you want to check their availability manually, feel free to access that segment of the device effortlessly to ensure you have not missed a notification.

To determine the current software version on your device, start from the Home screen

1. Select Settings.
2. Select System.
3. Select About.
4. Select Amazon Fire TV to Reveal the Current Software Version.

To manually check for an update, from the Home Screen:

1. Select Settings.
2. Select System.
3. Select About.
4. Select Check for System Update.

(If an update is available, it will automatically start downloading.)

Note: *After the download is complete, select Install System Update to install the update. The device will automatically install the next time your device is idle for 30 minutes or rebooted, if you do not choose to update immediately.*

How to Reset and/or Restart Amazon Fire TV

If you have a frozen screen, and need to restart your Amazon Fire TV, the solution is simple:

1. Press and hold the Power button for a full 20 seconds.
2. After 20 seconds, release the Power button.
3. Press the Power button again to restart your device.

If you would like to reset your Amazon Fire TV, begin at the Home screen:

1. Select Settings.
2. Select My Account.
3. Select Deregister.

How to Restore Amazon Fire TV

In order to restore your Amazon Fire TV to your settings, register the account using your Amazon information from the Home screen:

1. Select Settings.
2. Select My Account.
3. Select Register.
4. Enter the Correct Information for your Amazon Account.

How to Deregister an Amazon Account

1. Choose Settings, then "My Account" and then "Amazon Account."
2. Choose "Deregister" to remove your Amazon Fire TV from the account it is currently associated with.

Once you do this, you'll be presented with a "Registration" screen where you can choose to "Register" your current Amazon account, or "Create an Account."

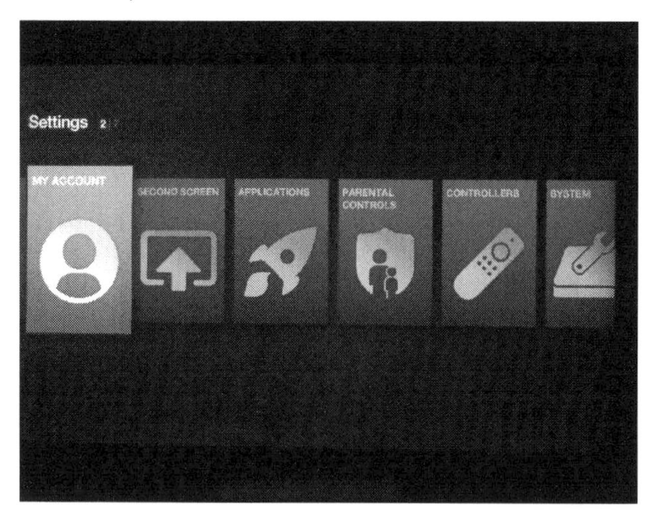

How to Register Amazon Account

1. Choose Settings, then "My Account" and then "Amazon Account."
2. Choose "Register" to register your account if you currently have an Amazon account. If not, choose "Create New Account."

You will be prompted to enter your Amazon account email address and password to register. An Amazon screen will display as it registers your account and then telling you your device is registered.

How to Use Amazon Cloud Drive

The Amazon Cloud Drive is an online storage account that allows you to store and access your personal photos and videos (as well as documents and music) online at any time.

With the free Amazon service, 5 gigabytes of storage space are provided free to all customers, with the ability to upgrade to more space. For example, $10 a year will get you 20 gigabytes of space, $25 a year will get you 50 GB of space, and $50 a year will get you 100 GB of space. You can actually upgrade to as much as 1,000 GB of space for $500 a year if you choose to use this as a way to store your various content. With the Cloud Drive App you can easily upload content from your mobile device to your Amazon Cloud account.

How to View a Slideshow of Photos

You can view a slideshow of your photos that are stored in your Amazon Cloud Drive.

1. Using the Cloud Drive, you can go into your photos or albums on your Fire TV by choosing "Photos" on the left side of the screen from your menu.
2. Next, choose the album you wish to view, or select "All."
3. You can choose an individual photo to view it on your screen by navigating to it and clicking the circular selection button on your remote.
4. Or, you can navigate down to where it says "Start Slideshow" and choose this option to see your photos displayed on screen.
5. Alternately, if you've got a photo up on the screen, you can click the play button on your Fire TV remote to begin a slideshow of your selected photos.

How to Set a Photo as Your Screensaver

You can easily choose an album of your photos on your Fire TV and use the photos as your screensaver.

The photos will continuously display when you've left your Fire TV idle for a certain amount of time.

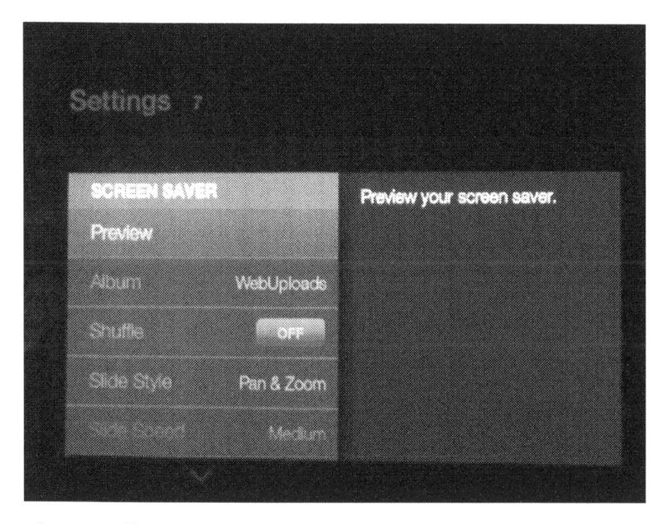

To set a photo album as your screensaver:

1. Go to "Photos" on the left side of your Fire TV's main menu and select it.
2. Navigate to the "Albums" area of your photos and select an available album of photos you'd like to use for the screensaver.
3. Next, on that album's page, navigate down to "Set as Screen Saver" and select it. You should see "Successfully set as Screen Saver" pop up briefly on your TV screen.
4. The photo album you've selected will display on your TV screen whenever you've left your Fire TV idle for a bit of time.

To adjust Slideshow settings, you can go into your Fire TV "Settings" at any time and go to the "System" option to select "Screen Saver."

This will bring up a variety of options for your screen saver such as:

- Preview - to see how your slideshow will display.
- Shuffle - You can set this "ON" to randomly display the photos in the album.

- Slide Style - You can set this to "Pan & Zoom," "Dissolve" or "Mosaic," based on how you'd like the slides to transition into each other.
- Slide Speed - You can set this to make the slide photos change faster or slower in the slideshow.
- Start Time - You can use this setting to adjust how long your Fire TV must be idle before the slideshow displays on your TV screen. The default time is 5 minutes. Other start time options include 10 minutes, 15 minutes, or Never.

Trouble Shooting Amazon Fire TV

Yes, the Amazon Fire TV is an electronic device. Yes, these devices sometimes have trouble. Here are a few tips to get you back on track, and enjoying your device fully.

If you are experiencing any of the following trouble with your device, like:

- TV Screen Fuzzy or Black
- Amazon Fire TV Not Responding
- Video with No Sound

Begin from the Home Screen of your device, and scroll to the System option. Locate the Display segment and scroll to reach the audio or video segment that is troubling your viewership.

Remote Not Working

If your remote is not working, there are a couple of things you can check first:

1. Check the batteries, or replace them with a fresh batch just to be safe.
2. Unplug the device fully, and plug it back in to reboot.
3. Starting from your Home Screen, Locate Controllers and Remotes, and follow the instructions on screen to re-sync or pair the remote.

Network Issues

If you are having network issues, it may be a matter of WIFI interference, which can easily be checked by accessing the "System" option from the Home Screen. Locate the WIFI section, which will take you to Network and WIFI setting and allow you to troubleshoot issues in this category right away.

Troubleshooting App Issues

You may find that a particular app (channel) you've installed on the Fire TV gives you issues. For example, as new apps become available from third party developers, some may be unstable in their first versions.

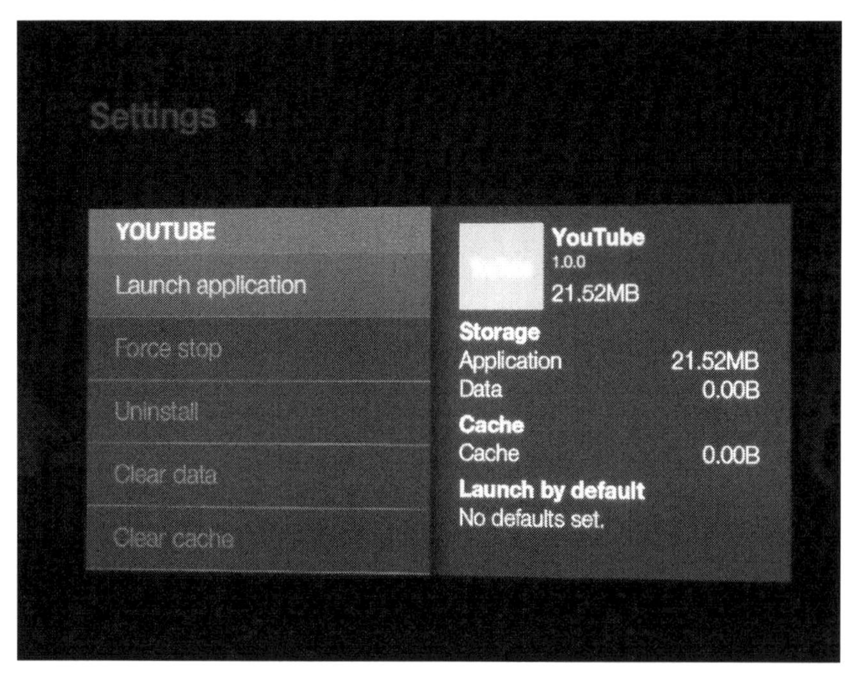

You can easily go into your Fire TV settings to troubleshoot and fix app issues.

1. Go to "Settings" on the left side menu of your Fire TV's home screen.
2. Choose "Applications."
3. You will now be able to scroll down through all of your installed apps.
4. Choose the particular app you are trying to troubleshoot.

From here, you now have various options including:

- Launch application (Used to immediately open that app on your Fire TV and use it.)
- Force Stop (This will force a running app to stop running on your device. It won't remove the app or any files)
- Uninstall (A last resort for non-working apps, you can use this to remove the app and all of its associated data. You may want to try this and then reinstalling the app to troubleshoot any problems.)
- Clear data (Proceed with caution before using this option as it will wipe out all of the data files, settings, accounts and databases you have associated with the app.)
- Clear cache (This option can be helpful for freeing up some memory or space on your device to help it run quicker.)

These additional options may help you troubleshoot issues on your device, should you find that the device is running too slow, or that a particular app is causing it issues. Keep in mind if you struggle with using technology, it may be best to have a more tech savvy friend or family member help you with these options if you run into trouble.

Amazon Fire TV accessories

Part of the allure of the Amazon Fire TV is the accessories that can be paired with it for your enjoyment. It is likely that new accessories will pop up on the market as the Fire TV progresses. For now, let's look at what is currently available.

Amazon Fire Game Controller

The Amazon Fire Game Controller is wireless and activated through Bluetooth technology, which means you can play all of your favorite games without a cord tethering you to the television.

The good news is, any Bluetooth Game Controller can be used (although Amazon is not touting that fact), and you simply activate from the Home Screen, selecting the Controllers option. The Bluetooth option will allow you to pair the controller to the device and begin playing games right away with the freedom you would enjoy with a game console.

There is also at least one third party Fire Game controller available. There is one from Nyko, and it is currently priced at $19.99.

HDMI and Optical Audio Cables

It is no secret that Amazon would like for you to purchase their HDMI cable and Optical Audio Cables, and with the first iteration of the Amazon Fire TV, it may not be a bad idea. You will need them - at least the HDMI cable, and since one does not come with the device, you will need it to install your new Amazon Fire TV. You can purchase the cable online at the same time you purchase the device, so they will arrive at the same time. Typically an HDMI cable costs around five to ten dollars.

As far as the Optical Audio Cables go, the idea is to simply connect your devices using these cables to enjoy a clearer and crisper sound from your device. Although it is not necessary for your device's use, the cords are available for around five dollars, so it may be worth the boost in sound quality.

Kindle Fire HDX Tablet

Yes, the Amazon Kindle Fire HDX tablet is an excellent accessory to have for the Amazon Fire TV. Clearly this tablet is a device on its own that has many capabilities in addition to being an accessory for the Amazon Fire TV, but it will work seamlessly with your new streaming media device and provide even more capabilities for the Fire TV.

The Kindle Fire HDX is a competitively priced tablet, and it comes in 7-inch and 8.9-inch models with 16 GB, 32 GB, or 64 GB of on board storage. You can choose the tablets that include "special offers," which are basically advertising, or you can pay a bit more, and receive a tablet free from these offers. You can also choose Wi-Fi only versions, or you can get models with Wi-Fi and 4G LTE from either AT&T or Verizon.

The price of the Kindle Fire HDX tablets ranges from $229 to $594 depending on the size of screen you choose, which "special offers" you choose, which amount of on board storage, and also which Wi-Fi and/or 4G LTE capabilities you choose. Even at its most expensive, the Kindle Fire HDX is an incredibly affordable tablet compared with tablets like the iPad Mini and iPad.

In addition to the Second Screen capabilities, the Kindle Fire HDX tablets feature an exclusive Mayday Button, which offers live video support at the push of a button straight from your device. This is a service that no other tablet offers.

Another great feature is the X-Ray for Movies and TV, which provides instant details about the TV or movies you are watching. Plus, X-Ray for music provides the lyrics as the song is playing, which can be incredibly helpful if you are trying to figure out a lyric. These features will enhance your TV, movie, and music experience with your Fire TV.

You can find out all about the Kindle Fire HDX in my best-selling Amazon book *Kindle Fire HDX & HD User's Guide Book: Unleash the Power of Your Tablet!*

Conclusion: The Future of Amazon Fire TV

As with most electronic devices, the Amazon Fire TV will only get better with time. The first iteration of almost all new devices is lacking in differing areas. Once each of the features is fully available, including the Amazon FreeTime, the device will come into its own as a fully functioning entertainment system for your home.

There is no doubt that the ability to play games using a Bluetooth controller puts this device above and beyond its competitors, but the overall use may require a few updates before it works optimally.

In the meantime, the ability to view videos, movies, television shows, and all of your Amazon content should work flawlessly, allowing you to enjoy a great movie any time from the comfort of your very own home. The device is available for $99, and is better than the Apple TV and Roku.

For more information about Fire TV and other technology tips, tricks, and updates, please visit TechMediaSource.com.

Other guides by Shelby Johnson

Apple TV User's Guide: Streaming Media Manual with Tips & Tricks

iPad Mini User's Guide: Simple Tips and Tricks to Unleash the Power of your Tablet!

iPhone 5 (5C & 5S) User's Manual: Tips and Tricks to Unleash the Power of Your Smartphone! (includes iOS 7)

Kindle Fire HDX & HD User's Guide Book: Unleash the Power of Your Tablet!

Facebook for Beginners: Navigating the Social Network

Kindle Paperwhite User's Manual: Guide to Enjoying your E-reader!

How to Get Rid of Cable TV & Save Money: Watch Digital TV & Live Stream Online Media

Chromecast User Manual: Guide to Stream to Your TV (w/Extra Tips & Tricks!)

Google Nexus 7 User's Manual: Tablet Guide Book with Tips & Tricks!

Roku User Manual Guide: Private Channels List, Tips & Tricks

Printed in Great Britain
by Amazon.co.uk, Ltd.,
Marston Gate.